Walid Kaakeh

Monitoring, Evaluation, and Learning (MEL) in Pest Management Projects

Walid Kaakeh

Monitoring, Evaluation, and Learning (MEL) in Pest Management Projects

Methodologies, Data Collection, Impact Assessment, KPIs, and Reporting

Noor Publishing

Imprint

Cover image: www.ingimage.com

Publisher:
Noor Publishing
is a trademark of
Dodo Books Indian Ocean Ltd. and OmniScriptum S.R.L publishing group

120 High Road, East Finchley, London, N2 9ED, United Kingdom
Str. Armeneasca 28/1, office 1, Chisinau MD-2012, Republic of Moldova, Europe
Managing Directors: Ieva Konstantinova, Victoria Ursu
info@omniscriptum.com

Printed at: see last page
ISBN: 978-620-8-87047-8

Monitoring, Evaluation, Research, and Learning (MERL) in Integrated Pest Management Projects

Methodologies, Data Collection, Impact Assessment, Key Performance Indicators, and Reporting

Prof. Dr. Walid Kaakeh
International IPM Consultant

First Edition – 2025

بسم الله الرحمن الرحيم

IN THE NAME OF ALLAH

THE MOST BLESSED, THE MOST MERCIFUL

Dedication

To my mother and father,
Aisha and Abdul Ghani Kaakeh
May God bless them and make them dwell in
His spacious gardens

To my wife Fatima Khatib

To my beloved children,
Tasneem, Yasmine, Yamen, and Enas

I dedicate this book

Walid Kaakeh

Table of Contents

1. Introduction to MERL in Integrated Pest Management

Monitoring, Evaluation, Research, and Learning (MERL) is integral in Integrated Pest Management Projects (IPM). MERL is defined as a systematic inquiry to increase understanding of and improve the effectiveness of policies and programs and the evaluation applied in such inquiry. It comprises: defining the boundary and scope of a project; designing and establishing systems, materials, tools, and training programs for the collection of monitoring and evaluation information; and using this information to improve project implementation. The projects aimed at preventing the opening and establishment of IPM-prohibitive 'pest pathways'. These will interlink to form an IPM research, capacity development, and implementation network across borders using new monitoring and evaluation processes, systems, tools, and training of the key stakeholder groups in four regions of Africa and Asia (Pretty & Pervez Bharucha, 2015).

The core components of the MERL are:

(1) Project Definition and Baseline;

(2) Process Monitoring, Evaluation, and Assessment;

(3) Research, Learning, Capacity Development, and Impact Assessment; and

(4) Reporting, Project Definition and Baseline.

Project Definition is a thorough, wide consultation with project stakeholders to elicit their expectations, understanding, and commitment to a successful project. This usually takes the form of semi-structured interviews with key informants, then a workshop to contextualize and finalize the project(s) including assumptions and rough project plans. Baseline conditions prior to project start are also assessed, and information on these collated. Baseline data can be obtained from databases or surveys. This forms the project scope, framing timebound project aims, and detailing preliminary input-by-output project activities (from which specific responsibilities can be allocated using Gant charts) (M. Matten et al., 2004).

Process Monitoring is the systematic collection and analysis of output-oriented indicators that are directly related to specific targeted behaviours. This practice serves to verify adherence to the established project plan. It involves the use of context-specific descriptions that encompass preliminary outputs, actual outputs, and ongoing outputs, such as dissemination materials, the learning salience of activities, and a broader political context surrounding the chosen issue or context. In addition to these factors, other valuable indicators may include the number, skills, and the extent of involvement of members of the Project Task Force. The information necessary for deciding on which behaviours were targeted, as well as who those behaviours were directed towards, is obtained from the project staff. Moreover, this information should be updated regularly throughout the life of the project to ensure accuracy and relevancy.

2. Methodologies for MERL

Successful outcomes and significant impacts of Integrated Pest Management (IPM) projects can be attributed to a multitude of interrelated factors that do not lend themselves easily to capture in hard scientific, quantitative formats, which often emphasize numbers and statistics. Nonetheless, Programme staff must be diligent in monitoring and thoroughly evaluating these projects to ensure that the investment made is a worthwhile one, both scientifically and socially, benefiting the communities involved. The challenges these evaluations represent are considerable and often complex. An exploration into how to effectively make and attend to such evaluations using participatory appraisal methods with great sensitivity and respect for local cultures may reveal promising paths forward in the evaluation process. To be socially sensitive and responsible, IPM evaluations should be participatory in nature, providing access to skilled interpreters and experts who can facilitate communication, while being careful not to impose potentially Eurocentric vocabularies of description upon the community members. When analyzing the evaluations' outputs, it is essential to do so through careful cultural interpretation, respecting the local context and perspectives.

The formulation of feasible solutions to agricultural challenges requires the active participation of those local individuals who are to carry the solutions out within their own communities. A noteworthy case study illustrating social sensitivity is 'participatory appraisal'—an approach that was pioneered by Robert Chambers and is specifically designed for use with local communities all over the world, helping them articulate their own insights and experiences. In a typical participatory appraisal interactive paper mapping session, for instance, farmers' diverse methods and their multifaceted effects would be visually drawn, probably in intricate detail, on a large sheet of paper. The interpretations and insights would be negotiated through a collaborative group analysis process, which would be visually tracked as writing on the bottom of the map, emphasizing and saturating key points that emerge from the discussion. As significant insights and deeper understanding emerge about why the project has performed as it has, it should be emphasized in the narrative description. It is crucial to ensure that these insights remain firmly rooted within the participants' own cultural forms of description. Moreover, community transformation in the IPM villages needs to be treated in its own terms and be interpreted contextually. If it is pushed outside these established cultural parameters, perhaps rephrased in dry words, intricate graphs, and cold numbers that adhere more closely to Northern scientific sensibilities, there is a substantial risk that the analytic insight and therapeutic understanding will be lost amidst the data.

Writing up evaluation outputs for the purpose of public consumption opens up the possibility of epistemological racism, parallel to how Northern scientific vocabularies have historically been used to deny legitimacy to indigenous ways of knowing about the environment and understanding nature. In summary, integrating hard, scientifically quantifiable analyses with soft, socially sensitive ones into an evaluation could ultimately serve as the holy grail that IPM expansionists seek, potentially allowing for a more comprehensive understanding and improved strategies in pest management practices globally. (Wienarto et al., 2016; Centre for Agricultural and Rural Cooperation, 2015)

2.1. Qualitative Methods

Integrated pest management (IPM) is recognized as a highly sustainable and effective approach to managing pests in various agricultural settings. This innovative method integrates multiple control technologies and diverse strategies aimed specifically at reducing pest populations in a balanced manner. The successful implementation of IPM requires considerable behavioral changes from key stakeholders such as farm workers, agricultural technicians, and farmers themselves. Adapting to these new practices can lead to significant improvements in pest management effectiveness. Furthermore, well-designed evaluation and learning systems are crucial in helping IPM projects thoroughly understand, assess, and ultimately improve both project performance and long-term effectiveness, as well as sustainability. Evaluation, ongoing monitoring, targeted research, and continuous learning are proven to be extremely beneficial for IPM initiatives. These practices provide essential resources, increase understanding of project designs, foster creative solutions and innovations, and enhance overall transparency and accountability within the projects.

Raising awareness about the latest monitoring, evaluation, research, and learning methodologies can significantly enhance the management and administration of IPM projects, leading to better pest control outcomes and healthier ecosystems. Emphasizing these approaches not only benefits individual projects but can also contribute to broader agricultural sustainability efforts across different regions. (B. Capegsan, 2019).

There is growing recognition of the importance of evaluation, monitoring, research, and learning in agricultural innovation systems, including integrated pest management (IPM) projects. AEI brings together a group of organizations and projects with common interest in the system-wide application of rigorous evaluation, be it qualitative and/or quantitative, to improve, and enhance IPM implementation, effectiveness, and sustainability. Monitoring, evaluation, research, and learning systems have, can, and do help face challenges in understanding, assessing, and improving IPM projects. Working on evaluation, monitoring, research, and learning can be beneficial for IPM projects. This includes additional resources, better understanding of project design, changed individuals, organizations, communities, societies, creativity enhancement, fostering learning, improved focus, better marketing, and improved transparency and accountability. There are opportunities for IPM projects regarding newer, especially qualitative and participatory monitoring, evaluation, research, and learning approaches (Hochachka & Thomson, 2009).

2.2. Quantitative Methods

This paper describes the Monitoring, Evaluation, Research, and Learning (MERL) activities that were undertaken as part Integrated Pest Management (IPM) projects evaluating in Africa, Asia, and the Pacific and how lessons will be shared ensuring wider uptake of policy and practice change. The projects specifically focus on the creation and support of farmer-led multistakeholder platforms whose mandate is to collaboratively evaluate at scale the impact of locally developed IPM technologies, reviewing the decision processes taken and their outcomes. Evidence of effectiveness has been shared with and on behalf of farmer led partnerships with governments, research partners, bi-national and international organizations, NGOs, and horticulture exporters partners, resulting in a range of policy and practice changes across countries. Details are given of why MERL was needed, what was done, findings to date, and lesson learned.

For the initial phase of the IPM Research for Development project in Africa, Asia, and the Pacific. Using two product examples, namely mushroom and biopesticide based IPM technologies and approaches, and building on the work by multiple working groups, farmer understanding of IPM technologies. The rapid increase in the conviction that the IPM practices will benefit both crop yield and the environment has led to increased investment in development and introduction of novel IPM technology and practices, particularly in Asia; region that is now considered the front line in battling both endemic and newly introduced diseases and pests.

The objective was to test and evaluate farmers understanding of and interaction with novel IPM technologies among two contrasting farmer target groups, namely rice growers in India and mushroom growers in Viet Nam. In all cases, the farmers were interviewed with semi-structured questionnaires. The questionnaire permit both qualitative as well as quantification of the degree of understanding; notably how farmers are interacting with IPM technologies and the key information sources. There were 4,080 farmers contacted using questionnaires in countries, especially in rice farming. Farmers understanding of IPM technologies in a biophysical environment dominated by the organophosphate pesticide and the increasing conviction of the organic principles are suffered based after the interviews on their peer to peer based understanding. Individual farmer understanding in mushroom farming were differed similarly to rice, but the intermediates understanding was rarely due to high-income involvement with novel technology.

2.3. Mixed-Methods Approaches

Mixed methods approaches are increasingly recognized for their ability to integrate both qualitative and quantitative methodologies effectively. Researchers frequently employ varied mixed methods to fully leverage the strengths inherent in each approach, while also compensating for the respective limitations associated with them. This blending of methods and methodologies signifies a significant paradigmatic shift within the prevailing methodological landscape, and it has the potential to liberalize established categories and variations in research design. However, concerns persist regarding whether the label of mixed methods is becoming unwieldy, particularly in light of the fact that researchers often engage in the simultaneous use of an assortment of methods. These concerns bring to light the uncertainties involved in clarifying roles and definitions within mixed methods research, which can contribute to a degree of confusion in the field. Moreover, the notion of considcring qualitativc and quantitative approaches in research without the explicit use of mixed methods has also drawn attention.

Evidence supporting the practice of mixing methods is often analyzed against various typologies that categorize mixing methodologies. In addition, these typologies seek to address the apparent confusions in how qualitative methodologies are defined, along with the ways such confusions are further compounded when claims of methodological mixing are made. It is important to acknowledge the skepticism surrounding the capabilities of researchers attempting to mix broad methodologies, drawing insights from the literature that serves as cautionary notes. In light of these considerations, a mixed-methods approach is frequently viewed as a preferable option for providing enhanced rigor, which ultimately leads to greater precision in the outcomes of research and a more nuanced interpretation of knowledge in its totality. Many observers regard mixed methods as the emergent paradigm of the twenty-first century, and their application is well-established across a diverse array of disciplines. Mixed methods are utilized in what can be considered some of the finest examples of both quantitative and qualitative research within various fields. Nevertheless, the current landscape of mixed methods literature is both burgeoning and scattered, which highlights the pressing need for a capable navigator within this area of study who can effectively capture the breadth and depth of mixed methods. This appreciation positions mixed methods as both a distinct field of inquiry and a collection of methodologies worthy of exploration.

In examining mixed methods resource articles, a data-driven heuristic emerges that grounds the evolution of this field, thereby making it more accessible to newcomers, as well as providing useful insights for specialists already engaged in mixed methods research. In this context, formal mixed methods can be defined as a practice that follows a pragmatic paradigm and adheres to a specific logic of inferences, which indicates that there is no monopolistic audience of readers. This fosters a quest for complementarity and encourages a broader understanding of mixed methods. Exploring the framework of analysis pertaining to mixed methods reveals three vital information blocs that are essential for understanding what constitutes mixed methods: epistemological assumptions, procedural considerations, and the necessary steps to ensure rigor in research. The intricate details of mixed methods, from the foundational epistemological assumptions to the critical steps for ensuring rigor, are discussed with the support of anecdotal illustrations. Furthermore, references to best practices and seminal studies from the global literature on mixed methods provide valuable perspectives that enrich the discourse around this methodology. (Almeida, 2018).

3. Data Collection Techniques

Surveys, Focus Group Discussions (FGDs), and Individual Interviews are the most widely used MERL tools to collect qualitative data. Surveys are particularly useful for the collection of quantitative data and key informant interviews are useful for collecting secondary data points on common practices, cultural context, and target population. However, literature on the use of these methods specifically in the context of Integrated Pest Management (IPM) is limited. This paper advances findings and recommendations from a scoping review of peer-reviewed and grey literature on MERL within Agricultural Development projects for key MERL stakeholders.

The scoping review included peer-reviewed publications and grey literature dated from 2000 to 2022. No geographical restrictions were applied. Publications were coded thematically, and findings were grouped according to the perspective of target users, which included Project Implementers (PIs), Monitoring, Evaluation, and Research Experts (MERL), Technical Experts (TEs), and Coordinating Agencies (CAs). Separate findings are presented here for each of these groups. Considerations that fall across multiple groups are addressed first, as are findings related to the design of this project's IPM MERL tools, and synthesis recommendations.

Between 2000 and 2020, the majority of agricultural development projects reported to CAs, donors, and/or funders with the tools and SMRs that had been developed at that time. However, few projects implemented self-designed tools and SMRs in 2020 and number of SCARPs fell by 83% from 2000-2020. This is concerning as SCARPs provide a nuanced analysis of existing systems. To access these agricultural development projects, one or more common indicators should be framed by PIs, CAs, and donors/funders. The tool to collect evidence and validate existing indicators could be developed first, and tools to collect additional data points for contextual understanding and evidence of other identified indicators should be developed thereafter. The purpose of these tools, data points, and indicators should be clear but limited in number to reduce the burden on TEs (Kosmowski et al., 2017).

The approaches from which qualitative data collection techniques can be selected include multiple levels of engagement, project types, research purpose, and dominant design approach. Tools such as survey questionnaires, checklists, and guidelines should be tailored for the local context. Primary data collection methods that should be emphasized include surveys, FGDs, individual interviews, and environmental observations. Common qualitative research sub-methods for these tools should be incorporated but prior training should be conducted in detail. As interviews carry high risk for bias, there is a need to assess prior knowledge and sensitivity of the respondents. Furthermore, the involvement of peers from the target groups in tool adaptation, implementation, and analysis would improve environmental and cultural case relevance (Cristina de Oliveira Lima et al., 2016).

3.1. Surveys and Questionnaires

Surveys are widely acknowledged as one of the most popular and effective means of systematically collecting MERL (Monitoring, Evaluation, Research, and Learning) metrics from IPM (Integrated Pest Management) project managers or their partners involved in various initiatives. They provide a relatively cost-effective and efficient way of gathering data from a large number of people, making them particularly appealing for organizations looking to gain insights without incurring significant expenses. When conducted online, surveys not only streamline the data collection process but also do so with minimal effort and in an automated manner, which enhances both efficiency and accuracy in the overall evaluation process. However, it is crucial to ensure that the survey questions and response categories are both relevant and exceptionally clear. This clarity is essential if the aim is to obtain useful answers from respondents and to minimize the risks of obtaining biased or misleading results, which can arise from poorly-worded or ambiguous questions. To enhance the effectiveness of surveys, it is advisable to carefully develop survey questions based on specific survey topics, and subsequently arrange those questions into well-structured surveys. These topics can encompass both project tasks and the outputs that result from these tasks. Understanding the distinction between these two types is vital; outputs' default questions should primarily focus on assessing the project's impact on the target system, whereas tasks' default questions should center around exploring the project's intentions, efforts, and methodologies. This discerning approach means that the resulting survey questions are necessarily nonstandard, reflecting the unique context of the projects being evaluated. Moreover, the response formats or categories for these survey questions must be thoughtfully designed to align with the meaning or intent of the questions posed. This alignment is critical in ensuring that the surveys produce meaningful and useful answers that can inform decision-making and improve future project implementations.

3.2. Interviews and Focus Groups

Interviews, both individual and in a focus group setting, are common methods of evaluation and are often implemented together. Focus groups are initiated, as questions are prepared and probe deeper thinking by prompting discussion among individuals. The facilitator and/or research partner guide the discussion and observe for tone, body language, and silence. Discussion is recorded or transcribed for later analysis and synthesis. Conversational interviews typically involve one-on-one sessions between a researcher and participant, but may also include small group settings. Due to the smaller degree of influence from paraprofessionals, the quality of response is often richer. Interviews with project participants can be tricky; it is challenging to analyze and synthesize cross cases due to the varying levels of knowledge and understanding about what happened in implementation. To counter this, common criteria for relevant questions can be established and prompts utilized to direct participants toward a common focus.

The staggered arrangement method, which involves research assistants interviewing key informants as an initial step before gradually engaging with less informed respondents, stands out as a particularly effective methodology. This approach balances the depth of knowledge and overall understanding of project implementation among the various interview respondents. Follow-up inquiries are absolutely essential, as they present a prime opportunity for research assistants to both ask supplementary questions and gather vital clarification on specific points. Paraprofessional implementation officers may also require clearly defined guidelines and a list of questions, ensuring they understand precisely what is being learned from the interviews conducted. It is crucial to prepare adequately for any potential debates and to engage in discussions around pre-existing understandings prior to the commencement of each interview. Furthermore, maintaining a level of flexibility regarding the venue and the timing of the interviews may prove to be pivotal, as environmental factors can significantly influence whether the interviewee or the facilitator is able to engage actively. More broadly, it is essential that all approaches to conducting interviews remain sensitive and respectful towards confidentiality concerns, ensuring the privacy of all participants is upheld throughout the research process. (Wienarto et al., 2016).

3.3. Field Observations

Ex-post assessments have sought to derive lessons learned through documented experiences of energy:image procedures from case study countries. In regards to monitoring, a mechanism to document at an early stage project impacts during the program was developed. In addition, pilot assessments were conducted from projects in various countries. The outputs from these pilot assessments help identify related strategies, products and initiatives, which are further elaborated in the following sections of the framework. The purpose of documenting lessons learned from energy:image products and processes was to employ this knowledge in providing guidance for new technology and training adoption projects. The case study assessments intended to describe how the products and processes were applied. Important is how the evaluation has assessed the products and procedures and how knowledge can be applied to other projects.

Consistent with a non-proscriptive approach, the case study templates provide several guiding questions for the assessment. Examined are qualitative procedures to facilitate assessments, which can be done with farmer groups or project staff alone on paper or integrated with interviews. Each case study comprises introduction, project description, products and processes, and overall assessment sections. An objective of the case studies is identifying results from conducting its products and procedures. An influencing factor for project continuity mentioned is relative interests of staff in the project. The delivery of training to local team facilitators was addressed in phases of project selection, project planning, project management, and project efficiency and effectiveness. Nevertheless, clear distinctions in phases were found more practical than originally conceived. Thus, case study report contents are broken down by project phase and addressed separately to provide insight into products needed. Also comprehensive to provide suggestions for data collection, analysis, presentation, and report. For the above-mentioned reasons, the assessment products were addressed together. A detailed project description and comprehensive timelines for data collection and analysis are evident in the case study reports. Both of these generated data through interviews, discussion groups, and observations. However, determining report contents from these was a difficult task as they needed to be comprehensive.

3.4. Remote Sensing Data

Remote sensing serves as a remarkably effective and low-cost tool that can deliver detailed visual interpretations of plant health, vigor, and stress across expansive fields, thus guiding decision-making processes in Integrated Pest Management (IPM). This technology enables the collection of remote sensing data from a variety of sources, including ground-based sensors, aerial platforms, or satellite systems, which can scan the fields at predetermined intervals. Additionally, specific field areas of interest can be imaged upon manager requests tailored to particular agricultural situations. The diverse sensor systems available offer various spectral ranges, providing a significant flexibility in selecting the appropriate sensors for specific monitoring applications that cater to the needs of different agricultural practices. Once captured, the remote images can be processed to produce customized indices or models that deliver direct and actionable benefits to the daily operations of managers' jobs. This processing is essential, as data interpretation is performed to convert raw information into meaningful vegetation indices. Following this initial stage, a different level of interpretation is required to produce high-quality intelligence information that can guide critical management decisions. The workload associated with these two interpretation levels exhibits notable variation, with delivering a high level of user-customized intelligence often necessitating more intensive processing efforts. Furthermore, a sound understanding of the specific objectives that need to be met within the context of the given agricultural scenario is crucial for the successful use of remote sensing data in practice.

A straightforward user-customized interpretation approach could consist of presenting the standard NDVI values to users first, and subsequently soliciting their further clarification regarding these default interpretations. This initial step can guide users to refine their understanding based on established benchmarks. On the other hand, a more intricate example can be illustrated through a specialized pest detection model that utilizes various spectral indices alongside an assessment of variable growth area determination. This also emphasizes the application of expert interpretation within the context of the remote sensing process. By closely monitoring plant geography and the vigor of growth, crop managers are afforded the opportunity to make timely adjustments to their management strategies prior to the onset of pest infestations. In addition, models for leaf area index offer significant insights that serve as a valuable barometer for directly timing routine checks for pest detection. The use of remote sensing technologies in agricultural practices has significantly broadened the scope of monitoring applications, allowing for commercial-scale utilization. Today, an array of empirical relationships and indices has become readily available to assist in the effective management of challenges related to weeds, nutrients, and water. These advancements have been made possible through the availability of cost-effective plank cameras and sensors that contribute to agricultural monitoring.

4. Impact Assessment in Integrated Pest Management

Integrated pest management (IPM) is a comprehensive approach that systematically incorporates all possible components and strategies needed to maintain pest populations at levels that do not inflict economic damage on agricultural crops. This multi-faceted strategy includes the integration of biological control methods, the use of resistant crop varieties, cultural control practices, and the careful and judicious application of pesticides when absolutely necessary. Although IPM has become the prevailing paradigm for crop protection on a global scale, its adoption has not seen the widespread success that many advocates had initially hoped for. Reflecting back to the mid-2000s, the IPM revolution was praised for its potential to significantly increase food production, alleviate poverty levels, and minimize the unacceptable risks associated with the livelihoods of impoverished farmers while simultaneously enhancing the environment. Now, a decade or more later, we find ourselves in a position where it is crucial to take stock of the current situation and make necessary adjustments to our approach. Around the world, it has been estimated that roughly 10 million smallholder farmers have integrated IPM methods into their practices. However, despite this adoption, the overall use of pesticides has not shown a corresponding decline. Additionally, there is a worrying absence of evidence indicating that the detrimental externalities associated with pesticide usage—such as environmental pollution, health risks to humans, crop losses due to phytotoxicity, and the development of pesticide resistance—have noticeably decreased.

One particularly telling example comes from Indonesia, where the average fruit and vegetable farmer has paradoxically escalated their pesticide applications while embracing IPM; their applications have risen from 9.7 to 14.7 per season over the span of a decade. This prompts the crucial question: why, after more than 50 years filled with inconsistent yet persistent investment in various initiatives such as pest population monitoring, research into bio-ecology, treatment trials, participatory action research with farmers, innovations in biological control, the development of resistant crop varieties, and boosting research capacity, is progress so stagnant? Unfortunately, the essential message of IPM has only evolved marginally since it first made an appearance in southern regions. In every instance, resources put toward IPM have been significantly reduced since it took on a global presence. What is most urgently required now is a fundamental recognition that IPM is not just a mere technological advancement akin to introducing a new crop variety, tank, tractor, insecticide, or irrigation pump. Rather, IPM represents a holistic approach to planning and organizational management. To rectify the issues encountered with a malfunctioning system, it is typically necessary to execute several interconnected operations concurrently. In the context of IPM, these actions might consist of conducting participatory diagnostics at the atoll or district level to assess current pest management and pesticide practices, accumulating essential knowledge from farmers regarding pest bio-ecology, choosing and experimenting with improved management techniques, and reinforcing farmer institutions alongside establishing connections with suppliers who can offer desirable and affordable pest management solutions, as well as unconventional advice that can further enhance pest management strategies and practices. (Pretty & Bharucha, 2015).

4.1. Defining Impact Metrics

Typical monitoring, evaluation, research, and learning (MERL) will assess project impacts in terms of crop production, incomes and yields per hectare, number of people using project interventions, and the resulting ecosystem service benefits. However, assessing these impacts might take several years, with longer-term impacts (but with proxies that can be assessed sooner to see change. The sustainable use of commercial bioproducts could benefit from elucidating the communication linkages running from awareness to supporting behavior. As the insecticidal value of biopesticides comes to light, integrated pest management projects intending to transfer their use to stakeholders will need to assess whether new treatments were simply introduced to farmers via extension materials from an IPM project, or whether genuine adoption has occurred. describe use as the sustainable, supported and durable use of a bioproduct which should result in benefits to ecosystem services, soil health, crop production and farmer incomes. In a similar way, extensions to fishing gear will need to examine the sustainability and durability of new practices, such as whether new hooks are managed more effectively than previous gear. A clear understanding of how, when and by whom equipment chains are acquired is crucial to ensuring that results can be thoroughly weeded (Kumar & Martin, 2017).

Wealthier farmers may have the capital to invest in improved gear, later adopting larger, more tenable boats. Similarly, triococcus could co-occur with the loss of the plant in the fields. What mechanisms are needed to ensure that facilitation funds are re-injected and effectively tracked to avoid capital leakage? Hypotheses on responding to subsidized trap uptake and maintenance in Madagascar (in light of the farmer-preferred design) will need to be re-thought if there is no local manufacture and no clear supply of replacement parts. What governance changes are needed to ensure that local institutions remain effective for management and monitoring? Clarifying these will enhance the relevance and sustainability of results (Centre for Agricultural and Rural Cooperation, 2015).

4.2. Longitudinal Studies

An important complement to the pastures and hydrology work described above are longitudinal studies of cropping systems in terms of soil fertility, pest, environmental, social, and economic outcomes, as well as environmental and gender impacts. From a land-use planning perspective, it is essential to not only assess soil fertility levels and trends at a given time but also to monitor soil fertility over a longer period to understand rates of change, both positive and negative, but also potential shocks that might occur as a result of extreme weather events. The design should be relatively simple, with sites chosen to represent four or five different trajectory types. The type or land use, and biophysical characteristics at the time of choosing the sites will be recorded, along with time since conversion, and plausible reasons for this. At least one plot will need to be selected in each trajectory type representative of either long-term continuous cropping and/or long-term pasturing. Time-since-pasture plots should already have data available on pH, C, bulk density, texture, permeability, perched watertables, subsurface drains and P levels in hydrologically vulnerable paddocks, machinable areas, and the bio-physical, economic, and social impacts of management change. It will be necessary to get access to the historical soil chemical dataset for custody transfer plus a small number of samples taken after 2005 to provide some suggestion of trends likely in the near future. The same soil samples have been analysed for N and S, and it would help prepare for shared understanding and expectations if interpretation were added to this data.

Long-term experiments with agronomists and soil scientists assessing management and environmental changes/additions at the site level going back to 1992 have been discussed with the care group and they are enthusiastic. There are a number of excellent ideas which will require further refining of which these are three that came to mind. Long-term effects on soil, biscuit wheat yield and water quality of diverse cropping systems and crop rotations as compared with continuous wheat systems, under contrasting P fertiliser application rates. Long-term effects on soil properties and water quality of perennial pastures or legumes in cropping systems. Whereas, soil chemical properties, sodicity, crop yield, and eurotrophic status of waterways entering the basin for wetland N removal systems were considered. Descriptions of trials, treatments, variables collected and samples taken are included. Management recommendations include use of two rotations alternated annually for the field "Crop & pasture straw" and "Legume pasture" inputs, ensuring mixing of straw from varied sources, and intensive monitoring of lime applications and crop response. Management requirements are well-described; analysis of water and soil samples undertaken, monitoring of crop yields, and fiscal modelling recommended as management planning and care group advice tools.

4.3. Case Studies

Creating responses tailored for specialised use cases that involve field-specific jargon and nomenclature presents a unique and challenging problem. When addressing target audiences such as medical or engineering students, as well as researchers, the queries may often revolve around complex and hard scientific topics within disciplines including medicine, dentistry, or civil engineering. The resulting challenges include the need to consistently meet stringent character counts, while simultaneously ensuring the proper degrees of complexity, nuance, and formality that are required for effective communication in each distinct field. It is essential not only to convey the necessary information but also to do so in a way that resonates with the specific backgrounds and expertise of the audience.

5. Key Performance Indicators (KPIs)

For the specific purpose of MERS, the following Key Performance Indicators (KPIs) have been meticulously defined to monitor and evaluate the progress of the project. These carefully selected KPIs track essential changes and validate the significant effort that is devoted to various research activities; several different MERS projects will specifically target these defined KPIs, and the respective outcomes and results of the projects will directly impact their overall value and effectiveness.

1. Pesticide sales (kg/ha)-A measure that tracks the quantity of pesticides sold. Project results are expected to increase the efficiency of farmers' pesticide use (fewer sprays and, hence, lower sales). It will take time to accurately

measure these indicators: they must be recorded systematically from shop keepers and/or producers of pesticides. This will represent a considerable effort over a large number of shops.

2. Aggregate pesticide shopping (k$CA)-Pesticide activity measured in dollar terms. The denominator will be local currency, and it will be adjusted for inflation according to available indexes. This measure tracks trends similar to the Kg/ha measure: the aim is that a cost-effective control will correspond to an overall decrease in sales.
3. Market price of pesticides (k$/ha)-Tracking the price in local currency terms. Reports of payments to farm-gate cooperatives will allow for accurate price measures.
4. Density of relevant Co-ops -Number of franchises per area or number of farmers per franchise. It has been established that the precise identification of co-ops will result from field work during phase II. Franchises will be geocoded and a measure of density will be derived for each village after calculating a projection surface. The aim of co-op policies is to increase the density of franchises through time, and hence increase competition and lower prices. Information on population of villages will be obtained from the national statistical office. This will allow for density measures in number of franchises per hectare or relative to the village population.

5.1. Selecting Relevant KPIs

Assessing progress towards achieving objectives for Integrated Pest Management (IPM) projects requires the selection of relevant Key Performance Indicators (KPIs), based on both and the type of changes being measured (Derr et al., 2005). This section identifies the most commonly employed impact, outcome, process, and contextual KPI types and provides guidance on selecting the most appropriate indicators within each category. The KPI selection process should be an iterative one. As understanding of a project increases, the most appropriate indicators may change. For this reason, indicator selection at project development and filling in this section can be prioritised on indicators that are well-defined and relevant to most projects. The KPIs that best measure a project's progress can be decided and ranked when considering a Best Estimate of I objectives' contribution to project outcomes.

Selecting KPIs for Integrated Pest Management (IPM) projects typically falls under six types – Context, Process, Output, Outcome, Impact, and other – plus subcategories within these types based on whether qualitative or quantitative changes are being referred to. For Integrated Pest Management Survey projects, categories differ slightly to better reflect common practices in survey projects. For scenario and modelling projects, additional "Computation" KPIs may be included that will not be applicable to most projects, which focus primarily on Surveys or Evidence types.

The assessment of each KPI involves determining whether it is common, assessable, and relevant to the project. Common assessments are those that are likely to be conducted anyway, such as national surveys or government reports, or those that are likely to be implementation and cost-friendly, such as simple questions on survey forms. Assessable assessment means the assessment can be done, as for some KPIs access to relevant data may be limited or countries may not conduct standard assessments/monitoring. Relevant assessments are those that are appropriate to measure a project's contributions to the agreed outcomes. For example, to assess the "Extent of IPM pest management technology use" indicator, it is necessary that IPM is defined prior to project commencement and this definition is within the control of the project to some extent.

5.2. Monitoring KPIs Over Time

Time-series data available over many years are paramount for stating whether trends and other variations are statistically significant. Also, factor analysis should rule out seasonal and calendar time confounding. Eventually, in the case of significant spillover effects, the identification of relevant within- and between-region group means is important (Boye et al., 2019). The estimates that corresponds to these generalization groups are the average spillover effects.

The joint estimation of the parameters of interest and the frailty distribution can be performed via the procedure "gllamm" (generalized linear Latent and mixed models), which can fit a wide range of hierarchical models within the general framework of multilevel modelling. This capability makes it easy to adapt the specification of the model. A pragmatic approach that allows to maintain close contact with the substance of the problem by starting with the simplest version of the model that uses aggregate data, and then gradually results to more complex and disaggregate models (Matten et al., 2004). This strategy facilitates the task of interpreting results and reduces the risk of an unfortunate specification of the data allocation and link functions in more complex models.

Further challenges arise when information from policy evaluations performed in different contexts are pooled in the same model, even if the research questions are equivalent. Different generations of the questionnaire, non-standard ways to translate the questions, varied contexts, different time periods for data gathering, dissimilarities in the design of the data collection process, or scarcity of data from specific groups, such as policy makers, can contribute to inconsistency in pool datasets. Pool analysis can handle these challenges and provide valuable hints and insights. As a general principle, comparative analysis should take as its starting point models estimated using one and then the others of the data sources, in order not to lose in precision but attempt to extend knowledge by analyzing the differences, in terms of estimates and significance.

5.3. Adapting KPIs to Local Contexts

Global awards for Integrated pest management (IPM) projects are primarily monetary, in the form of Requests for Proposals (RFPs) that outline the project funder's interests or priorities. These are generally, though not always, framed by the expected outcomes for IPM projects (i.e. pesticide reduction, diversity of IPM practices adopted, literacy improvement, etc.). Once these broad indicators or KPIs are framed by the technical working group of the funder, key performance indicators (KPIs) require technical scrutiny to ensure that the indicators proposed for evaluating and monitoring the IPM project's interventions can be tracked in the given conditions.

Technical consideration in selecting KPIs includes matching practices with project goals, issues with direct observations and thresholds, availability of data and information, plus challenges posed by local conditions. In cases where local conditions dictate substantial changes in KPIs (as is often the case), recommendations for revisions must be accompanied by such thorough detailing that the funders can be guided in planning for the prudent acceptance or revision of the indices, evaluations, and other measures of project success (Muleme et al., 2017).

In addition to the conditions cited in the "Technical Events" section, there are other important aspects of local contexts that the KM component must take into consideration in advising on the selection of indicators. Beyond the technical feasibility, a KPI must also be socially and politically acceptable if it is to be appropriate or prudent. While it might be desirable for a KPI to comprehensively monitor processes, some may require sensitive data that, while useful, are too sensitive for reliable collection. Social respectability and social desirability bias are applicable to the setting of KPIs for Malaysia as they are for any location. To be cautious of socially undesirable behaviors, stakeholders in Malaysia may be reluctant to choose indicators that would require sensitive self-reported data on issues such as sexual practices or pesticide price fixing. Stakeholders may even resist adopting indicators that require observation of indicators that, given current behaviors, would require leading questions. In settings such as these, it is imperative that indicator selection accompany thorough stakeholder consultations and community engagement through face-to-face workshops and online questionnaires.

6. Data Analysis and Interpretation

The analysis and interpretation of quantitative data is conducted either using Excel or SPSS software. At the start of all analyses, the assumptions of each technique should be checked. The minimum/maximum values of each variable should be checked to ensure they do not exceed the limits set in the questionnaire and raw data. A percentage of missing data should be computed for each variable and analyzed according to pre-defined rules. The reliability of instruments and scales should be checked using Cronbach's alpha, where values of less than 0.500 indicate that the scale is not appropriate for use. After data cleaning, hypotheses are analyzed according to the type of questions asked in the questionnaires. Social-demographic data such as sex, age, education group and wealth group of all respondents are analyzed descriptively. The validity of the intervention as perceived by respondents, satisfaction of the intervention, and benefits/results perspectives of the intervention are analyzed using analysis of variance (ANOVA), where hypotheses are rejected when $p<0.05$.

Analysis of qualitative data involves a different approach. Respondents are asked to quote verbatim their responses to open questions where some responses may be very lengthy in sentences or paragraphs. Comments are provided where relevant and some narrative details are given regarding the open-ended questions posed to all respondents. Direct quotations are used to illustrate some points. A framework or concept map is developed to categorize responses around themes and in include respondents' thoughts to substantiate them (Muriithi et al., 2016). The details in qualitative are coded and simple narratives are produced around the themes/categories of words. Narrative analysis is also used to explore the structural properties of text and others (Star, 1999). Text is taken as speech, where the researcher looks how something is said rather than the content. Interpreting prose text is a closely-reading process that focuses on specific words, expression and meanings. The tone, pauses, inflections, body language and gestures are equally interesting as stories are being told or read.

6.1. Statistical Analysis Techniques

The two examples illustrate practical, analyzable projects around fishable and swimmable waters in river basins that take full advantage of opportunities for statistical analyses of a broad class of observations, while incorporating knowledge of other potential effects where feasible. For fishable waters, where demographics are included, a national overview with a simple dose–response model makes it easy to see large associations with larger urban areas and smaller positive trends for those areas in recent years, with other analyses sharpening those points.

For swimmable waters, cross-basin and within-basin correlations show that some uses are correlated with beach advisories of greater duration and diversity in many basins, nicer beaches are correlated with fewer advisories; yet some basins differ markedly from those overall patterns, suggesting that extensive modelling could be much more informative if it did not easily mask such findings (B. Barr et al., 2006). Coastally located basins, industrialized basins, and those that are large, largely agricultural, with a high population density or with Zebra mussels tend to have fewer advisories. It may take years for teams to identify all available observations, but with standard treatments observations from a wide range of years can be compared effectively, and baths, without in-person visits to all sites, can be targeted for acquisition of more localized data to compare fish populations downstream of large vs. smaller treatment plants. In many practical applications, this hierarchy of linear model analyses will yield very useful results well within those deadlines. Other analyses require simulation, and can be even more informative but tend to be quite slow, as simple cross-basin plot on simulated observations illustrates. Though more time-intensive, research teams have known the simple cross-basin scatterplots to be the best exploratory analyses because each point represents an entire basin, an important political jurisdiction, as thoughts on basins are voiced. Even without statistical analyses and with just visual plots, the comparison results remain insightful, such as: as expected, total loads to ambient loads fall off rapidly; a mountain versus coastal separation is clear; but within each region, E. coli advice has not kept pace with dramatic E. coli increases; rains make things worse, and the future promises to aggravate lakes' E. coli problems. Recognition of the improvements achieved internationally in fish consumption advisory projects has led to new statistical expansions of the initial analyses.

6.2. Data Visualization Methods

A wide range of methods can be applied to visualize your data and make it stand out from across projects. Microsoft PowerPoint is a great tool to work with during on-ground presentations. This software has a lot of built-in utilities and flexibility for creating visuals. It has a good user interface and allows for creating visuals from scratch or using templates that are already present on the software itself or online (Lyman, 2019).

Charts and Images: Prepare your outputs with graphics. Choose bar or pie charts for the comparisons, number-line scale visuals/dials for indicators or trends, and annotations with text boxes, arrows, and shapes for important points that deserve highlighting. Separate these charts from the clutter of tables or grids to make them visible. However, talk about the charts or images when presenting since the audience may not read those notes and it is very important to understand the message. This is also the reason why the text should be avoided on visuals at all times.

Table: Choose to visualize your data with a grid and fill in the cells with colors based on the nature of the performance if the point is to stress the results and give a general assessment, that is a comparison of performances. But write down the outputs in a separate table for discussion as a first point since columns with colors will not present the actual outputs or values and they lose its power.

Maps: Maps with legends can be used on the other hand to share the results on the projects with the outside stakeholders, mainly funding and donor agencies. Good maps with colors, legends, and arrows/tags can easily get the audience's attention since it is taking place on their level of work and authorities. Therefore, the maps should be the first visuals to be put. But texts should be avoided at all times on maps since the crowded visual will confuse the audience instead of facilitating the understanding of the message. Instead, it is better to annotate the maps when presenting.

6.3. Interpreting Results for Stakeholders

The question of who a Monitoring and Evaluation (M&E) system is intended to serve may seem trivial or obvious. After all, the stakeholders who design and oversee a systematic M&E are usually from a particular organization or agency. However, like vulnerable youth who may feel similarly excluded from larger systems of governance, stakeholders also have different degrees of power over the M&E system. Therefore, there is no universally agreed-upon meaning of stakeholder, nor a single way to identify those affected by the M&E process and therefore able to develop or construct it (Sokol-Oxman, 2015).

The importance of stakeholder participation is generally held to be self-evident (despite the diversity of stakeholder perceptions). Participatory M&E is widely claimed to empower marginalized citizens and less powerful actors by involving them in the construction - and therefore the content - of evaluation communication. Modern participatory evaluation advocates go further: when the M&E system is genuinely developed and run by stakeholders, there is less danger of the dominant culture or worldview being represented. This distinction may be useful in developing a conceptual framework to explain the different levels of stakeholder involvement in project M&E. There are three categories of participation: stakeholding accounts, systemic involvement, and project involvement.

Stakeholding accounts refer to the general interests of stakeholders in the project. Stakeholders hold a stake in any number of senses: they are indirectly associated with the project, are likely to be affected by it, or even oppose it. An example of this type of M&E exists in China, where officials are responsible for communicating the impacts of a project organization-wide and do not have any review authority over the impact communication process. By contrast, local stakeholders with systemic involvement in development project projects and implementation are directly associated with M&E processes. Here, stakeholders perceive stake in multiple dimensions, including project design, expected impacts, or benefits. In this case, they may also be called participants. There is a distinction to be made on this term as well: different stakeholders may have different degrees of systemic involvement, based on their power to affect project implementation.

7. Reporting and Dissemination of Findings

7.1. Creating Comprehensive Reports

A fundamental component of MERL approaches is the ability to generate comprehensive reports, which involve the synthesis of project monitoring and evaluation information. Reports can be used for various purposes, including project planning, performance analysis, troubleshooting on concerns, or even staff capacity-building. As a result, reports can come in different formats or lengths, depending on those intended purposes. MEAL framework, guidelines, and other documentation are some of the formats used for reporting at the design and onboarding phase of a project. In addition, there are periodic documentation formats, which can vary in frequency or disclosure levels.

The periodic documentation of project reports can occur on various bases, including daily, weekly, monthly, quarterly, bi-annually, and every five years. There are also reporting formats classified by disclosure levels. For example, some reports, such as the base-set monitoring reports prepared at the project onboarding phase, are not disclosed beyond the project team themselves. In contrast, some reports, such as program-level impact reports, discrepancy analysis reports, and baseline fund reports, are highly likely to be disseminated at the donor level and widely publicized at the organization level, alongside avenues such as media outreach, policy frameworks, and interventions.

When generating reports, it is essential to explore the context, audience, and timing of the reports in advance, to proactively mitigate the risk of producing reports that will not be read. Reports targeting different interests, such as findings, recommendations, and risk management, should be prepared and finalized well in advance of their respective due dates. Reports are typically finalized 3-7 days prior to their due dates, and late-generation and distribution of reports is one of the sanitary defaults of reporting. Nonetheless, lack of knowledge of the audience, timing, and context of the reports could sometimes lead to greater doubts concerning the performance of the project as a whole and, as a result, distrust from wide-ranging stakeholders.

Typically, only formally approved terms of reference can initiate a co-writing process. The need for reports, as well as discussions on the audience, context, timing, format, and style of reports, should generally lead the draft report to be produced by MERL specialists rather than the experience of the concerned aspect. Chiefly due to this, most reports are drafted by MERL specialists.

7.2. Utilizing Digital Platforms for Dissemination

The promotion of Integrated Pest Management (IPM) technologies must reach out to users and stakeholders before they may be fully adopted. This dissemination or outreach process spreads technologies from researchers, extension workers, and agents to users, in an attempt to persuade users to adopt these technologies. Outreach may start with different adopters and use different extension agents; thus, sustainability depends on the local context and needs. Outreach may also develop local innovations and make new demands on research and development. Therefore, the desired stance for this part of the paper is as follows: first, to detail the demand for outreach, and second, to describe the philosophy and strategy of outreach actions. Illustrations are drawn from experiences in IPM eradication projects (Centre for Agricultural and Rural Cooperation, 2015).

A note on terminology: Dissemination is a general term that subsumes outreach and follow-up. It may refer to dissemination at higher levels of technology generation, such as radio broadcasts or technology fairs. Outreach, also called dissemination, refers to actions in the field of the target population or specific stakeholder groups (e.g., decision or policy makers, institutions). While this part of the paper centers on outreach, it acknowledges that there may need to be greater attention and action on dissemination. Monitoring and evaluation (M&E) also has a broader definition when intrinsic and extrinsic at higher levels of technology generation aspects are also considered. However, this part of the paper concentrates only on the M&E of outreach actions. Within this scope, M&E is understood as follow-up actions after outreach and may assess two types of impacts: impacts on the ticket population and effects on the technology itself.

Outreach actions must be monitored, evaluated, and followed up to determine the extent to which they were effective in bringing about adoption of targeted IPM technologies. The extent of adoption is assessed against the outreach strategy or plan, especially through identifying any gaps and/or shortcomings. In-depth follow-ups, such as case studies, may need to be undertaken and feedback given to relevant stakeholders such as researchers, extension agents, and policy-makers to fine-tune their roles and inputs.

7.3. Engaging Stakeholders in the Reporting Process

No project or initiative exists in a vacuum; each operates in the context of a complex interdependent environment. The conditions that enable or hinder a project or initiative are at least as important as the effort put into the planning and implementation of it. Engaging stakeholders (people or groups with a vested interest in the project) in this engagement process at every stage is essential because they are the ones who know the system best. This approach requires different reporting methods and modes of engagement, depending on the audience. Engaging stakeholders in planning, implementing, and analyzing systems thinking can lead to more nuanced recommendations for all project stakeholders.

Most integrated pest management (IPM) projects consider stakeholder engagement in the design phase, e.g., involving farmers in the design of monitoring sites/plots. Beyond this, best practices for ensuring stakeholder engagement throughout the rest of the project life cycle, including reporting, have not been broadly disseminated within agricultural research initiatives. A key tool for ensuring stakeholder engagement in reporting is the stakeholder engagement framework. Stakeholder engagement frameworks can help ensure that relevant stakeholders are involved in each step of the reporting process, from formulating questions and planning analyses to disseminating results (which may require a different mode of communication from IPM practitioner to funder).

The stakeholder engagement framework gathers information on the project scope, purpose, and timeline. During an initial brainstorming session, the project team identifies the key reporting steps in three phases: project initiation, project implementation, and project completion. Projects may develop plans for some reporting steps which allow only limited stakeholder engagement, but the project team may still want to keep track of how the stakeholder engagement framework is to be employed in other steps. The project team then engages stakeholder representatives in a series of meetings to refine the reporting steps while ensuring that relevant stakeholders are consulted. The stakeholder engagement framework was developed based on a workshop conducted in northeast India that brought together researchers, extension workers, and farmers to analyze results from a benchmarking exercise on varietal performance of lowland rice (Sokol-Oxman, 2015).

8. Challenges in MERL Implementation

As Integrated Pest Management (IPM) is having multifaceted components and roles, it is necessary to develop a comprehensive framework to evaluate it. In summary, performance indicators for assessing the progress of the 5Rs were proposed. This is not a mere box-ticking exercise; they were designed to examine how MERL processes and systems help different components of IPM work effectively. For the outcome and performance indicators of the outcome goal, participatory systems thinking approaches were favoured, and the use of people-centred visual evaluation tools was suggested to increase farmers' ownership of the process (Schuetz et al., 2017).

Measuring agreements, contributions, and effects after the direct participatory processes are more difficult in systems-oriented approaches. Optional additional follow-up attempts can be considered in preparation for these aspects. A strategic communication plan based on good practice principles is needed for effective knowledge transfer and one-off MEL activities. To enhance the internal sustainability of MEL, continuous capacity development has to integrate management and development by keeping a balance between social, environmental, and economic aspects. Adequate time, expertise, and understanding of the local situation are necessary for MEL development, as well as open communication and building trust and relations with the local people and other stakeholders.

As IPM is a complex adaptive social, economic, technical, and ecological system, there is a high level of uncertainty concerning the effectiveness and efficiency of solutions. In addition to knowing what type of MEL is effective, the use of diverse MEL approaches and indicators in combination, sensitivity of assemblages of indicators for changes and their context specificity also require further exploration. Multiple and fuzzy science–policy–practice interface networks favour knowledge co-production adaptive management and assessment approaches to respond to local knowledge needs in changing circumstances: Further explorations in this area are required. The implementation of the 5Rs by local facilitators and subsequent systematic observation and feedback will need further exploration in relation to research for a better understanding of how these approaches can harvest collective intelligence. Validation at national and international scales also requires further attention (Parsa et al., 2014).

8.1. Resource Constraints

Resource constraints affecting research, innovation, and dissemination efforts are relevant both to integrated pest management (IPM) programs generally and those specifically targeting a particular crop (Parsa et al., 2014). The research effort may be impeded by lack of appropriate tools and methodologies, lack of sufficient in-house personnel with the skills and training to carry out research, and lack of funds. There may be a dearth of research organizations that can be brought in to assist. Translation, adaptation, and dissemination of research results into working IPM packages may be unfeasible. There may be a lack of personnel who can be trained to carry out IPM training or lack of personnel and materials to carry out mass media training. There may also be difficulties in utilizing ongoing, large-scale crop production enterprises to widespread application of training packages. Ideas for IPM packages for the target crop may be scarce, and lack of knowledge of IPM practices in other climates or settings may limit the adaptation of these packages.

Involving pest management clientele in the continuous-looking phase of a project could allow understanding of resource constraints before embarking on a potentially fruitless exercise. Participants from states with no targeted crop may have difficulty appreciating why researchers and extension personnel from elsewhere should be involved in the project. It is possible that local pest management clientele may either not appreciate or not have any knowledge of how consultation with outsiders can ameliorate the situation. If a co-champion is a stakeholder with the resources and stature, a request to team up with outsiders may not be forthcoming. There may be little faith in the ability of outsiders to assist when they have no experience in pest management for the target crop in the area. Furthermore, there may be difficulties in involved personnel gaining understanding of the farming economy and agroeconomic environment of the area. Lack of understanding may lead to inappropriate or ill-fitting information and training packages being produced.

8.2. Data Quality Issues

Data quality is a complex and multifaceted issue that encompasses accuracy, reliability, validity, representability, precision, completeness, accessibility, timeliness, compatibility, and consistency. These are generally categorized into three to five high-level issues, including accuracy and credibility; reliability and interpretability; and representativeness, timeliness, and compatibility (Yang et al., 2013). Many high-level issues are related to each other. For instance, representability is related to compatibility, and credibility is connected to accuracy and reliability.

Although various data quality issues and sub-issues are reported in theory, it is usually difficult to apply them. For example, students' performance data is recorded in a report form at the end of each semester in one institution, while another institution records them by collecting performance in standardized tests during each semester week. Both institutions ensure accuracy and reliability. However, inability to access the data and difficulties in accounting for differences in data recording, population, and testing conditions means that the two data sets are not compatible. When addressing such data quality issues in practice, it is necessary to explore the relationships with practical applications and to design appropriate user-friendly visual tools.

Relating data quality issues to practical applications promotes awareness of the importance of data quality issues and fosters attempts to address certain issues. For instance, data users are often concerned about data accuracy and credibility, so data origin and processing procedures need to be carefully documented. Similarly, in data fusion applications, users are often concerned about data compatibility, so much effort should be devoted to providing compatibility checks and harmonization tools. On the other hand, the technical feasibility of checking, enhancing, and visualizing quality issues is an influential factor.

8.3. Stakeholder Engagement Barriers

Australian snout beetle (ANB) levels were monitored at two locations for two years in partnership with residents and non-profit advocacy organizations. Cooperative extension agents are tackling local food safety issues and the introduction of new pathogens. Provident materials are needed for some non-native species. There is a need for more research on species predicted to become pests as a result of climate change. There is increased inspection and regulation of interstate ornamental plant propagators to prevent the unintentional introduction of new pests (Parsa et al., 2014). Barriers exist at the social, institutional, and political levels in research and extension organizations and universities. Both trained personnel and funding have been cut. Academics have become more specialized and less broadly trained in disciplines that support a systems approach, which has contributed to the perception of unable to provide relevant competitive applications. New paradigms for research, funding, and extension that include indigenous scientists, community research with and for the underserved, the funding of assessed needs, and partnership with stakeholders need to be pursued, along with the acceptance of culture with and by stakeholders.

9. Best Practices in MERL

The Integrated Pest Management (IPM) Research for Development (R4D) projects that end their funding cycle, yet have a continued need to monitor and evaluate their impact, face the potential loss of much valuable data and evidence about their contributions to global food security and rural livelihoods. In this section best practices are outlined that take into account the unique characteristics of integrated pest management, the challenges of monitoring and evaluation as well as the need for longevity. The best practices integrate quantitative and qualitative evidence, recognize the challenge of interpreting data from projects that were active in multiple countries for a decade or longer, involve both stakeholders who were previously involved with the projects and those who were uninvolved, and involve universities, NGOs or other organizations who can act as custodians of the data and evidence (Schuetz et al., 2017).

Integrated Pest Management is a mix of diverse interventions that often vary remarkably from environment to environment, in practices as well as in the response rates of the pests and the surrounding ecosystem. Large-scale projects are therefore, by definition, not aimed at developing an all-encompassing package of IPM interventions. Rather, much of the knowledge and data generated lies with the collaboration and diversity of partners and funders involved to satisfy a multitude of contradictory variables. This heterogeneity presents a challenge in terms of measurement and interpretation. Though changes in farm practices can often be quantifiable in terms of impact on pest management, documentation of changes takes the form of many different qualitative interpretations rendering future objective measurement impossible.

Evaluation of large international development projects takes place several years after the completion of the project and is conducted through review of project documents, impact evaluations, and stakeholder consultations. In the case of large international projects, stakeholders may include the high-level results framework development team, which may be based in a different country from where the project has taken place, rendering consultation difficult and leading to such distinguished groups as local community members, officials, management staff, and academics each potentially providing their own interpretation of project impacts. This groups' set may also include new stakeholders who are unaware about the program and who can provide an external viewpoint on the measurements and sources of evidence of performance.

9.1. Integrating MERL into Project Design

Specific questions of an evaluation will be based on the objectives of the evaluation. The objectives may include assessing full or partial outcomes of the project, gathering evidence to support evaluations of success or failure, ensuring accountability through verifiable evidence, understanding how and why something happened to document lessons learned and inform future programming, ensuring that the program continues to be relevant and addresses changing needs, or providing evidence to inform the drafting of further proposals (Clark & Apgar, 2019). A distinction is often made between formative, real-time, process and impact evaluations. There is a further distinction made between process evaluation, evaluation of progress towards agreed milestones, and impact evaluations that assess the extent to which outcomes can be attributed to the project interventions. Some evaluation questions may not be answerable due to constraints, and it is important to be realistic about the scope of the evaluation and the questions that can be answered. Some questions may be inherently difficult to answer, such as what predicts success or failure at the full outcome level.

Given the timeline and budget constraints, the security risks, and to ensure independence, the evaluation is likely to be contracted out to an external expert and comprise several consultants, with some members of the co-design team able to support the process. The phase-in innovative programming and methodology is similarly getting underway, with work being done in one country prior to full implementation across the regional program the following year. Eating in evaluation approaches may mean that there is no or little baseline (or the data capturing what changed may not be comparable or is not widely available). The lesson learned, review, reassess and implementation process may not occur before the decision deadline. Evaluations are not usually only one-off, but rather occur as repeated (or continuous) assessments over time, focused on specific criteria or dimensions. Further reviews may be just light-touch assessments, target change in focus countries or centers of excellence, or take on new science themes.

9.2. Continuous Learning and Adaptation

Learning is an equally important but often overlooked aspect of monitoring, evaluation, research, and learning (MERL) for integrated pest management (IPM) projects. Organizations need to assure those aspects for themselves and for projects, particularly for multi-partner, multi-organization, long-term projects like those operated by the International Agricultural Research Centers (IARCs) that comprise the Consultative Group on International Agricultural Research (CGIAR). Non-project members had little insight into the project's actual operations or the impacts of the project until they were invited to participate in a systematic retrospective review. IARC MEL assessments would do more than examine what was done and what happened; they would explicitly and deeply examine how things were done to help design subsequent actions and influence future proposals. Some mechanisms were purposefully inclusive, but others were considered and claimed to be inclusive while being strictly limited to covering the interest of the project only. In many projects, the already loud voices of powerful groups became louder in the reviews. There was capacity for some voices to be heard, but inadequate willingness or commitment to do so.

Many project mechanisms to assure or identify lessons learned were in place and fully functioning, but the analyses of what had been learned were neglected. All projects had formal and credible MERL frameworks. However, MELs were narrowly focused on ensuring compliance. Learning from the analysis of MERL findings was almost entirely absent; lessons were rarely made available to the wider community that funded or cooperated in the project. Transparency was again highly variable, with one project openly sharing benefits with others and another requiring confidentiality agreements to avoid release. Many lessons were addressed in a cursory fashion and focused on how to structure a valid research project rather than depth issues of developmental approaches and equity.

Knowing what has happened to a project is important, but knowing that actors are seriously analyzing impacts and planning on how to improve matters is essential. A new project ideally needs a collaborative system of MEL tools and strategies to highlight how assumptions and expectations are being or are not being tested, and for whom, with a clear culture of action that engages all stakeholders (M. Avila et al., 2016).

9.3. Building Capacity for MERL

Capacity as a process for developing knowledge, skills, and attitudes that enable behaviour change among people, organizations, and systems has emerged as a dominant approach to guiding Public Health programming in developing nations. Capacity building takes place through training, mentoring, experience, and knowledge sharing. These processes lead to a better understanding, awareness, and development of systems, work processes, or organisations; reflect values such as collaboration; and involve the hard and soft aspects of development such as funding or management processes. Importantly, capacity building/ interventions/capacity strengthening can thus take many different shapes and be implemented in many different ways. The phrase "to build capacity" can be used to define the action, but "capacity development" represents the more complex understanding of the process (Garley et al., 2016). Building capacity for MERL or capacity development for MERL would include efforts made to build knowledge, skills, ability, or willingness to undertake the systematic collection, analysis, interpretation, and use of information to understand and improve data use in Integrated Pest Management projects. This understanding can be further elaborated into four distinct areas of MERL capacity.

Overcoming obstacles to quality MERL generally means working, or collaborating, with institutions or organizations directly responsible for MERL.

Depending on the context, this may include local or national monitoring and evaluation offices and practitioners, international agencies and bilateral aid organizations, research centres, universities, and institutions that actively use project results in designing strategies/policies or informing the general public. From those corners, expert knowledge and best practices in MERL can be shared and adopted. To reduce bias in information and talent pools, MERL players should ideally be recruited from multiple sources and backgrounds. The goal is to enable the systematic collection, analysis, interpretation and feedback of results at all stages of project cycles/interventions, and to strengthen the role of feedback and learning in policy design and project accountability.

10. Case Studies of Successful MERL in IPM Projects

The Asia-Pacific region faces significant threats to food security which this writing classifies as dynamic challenges and chronic challenges. Dynamic challenges encompass conflicting trends such as population growth and declining agricultural labor force which are driving increased demand for food yet limiting food production capacity. They include the emergence of new livestock diseases, new varieties of insects and weeds that affect crops, and changing consumer preferences. Chronic challenges include poorly aligned incentives among agricultural stakeholders and insufficient national and regional mechanisms to address transboundary agricultural problems. Recently initiated, the ACIAR (Australian Centre for International Agricultural Research) research investment 'Integrated Systems for Managing Pests and Diseases of Agriculture in the Asia-Pacific Region' aims to meet these challenges through a participatory approach involving a range of stakeholders including farmers, farmer organizations, agribusinesses, government departments, research organizations and universities. The 'integration' in this research investment refers to the collaboration of stakeholders across these three domains and the integration of knowledge and practical experiences of disparate stakeholders to enable informed negotiation of choices that improve pest and disease management and research decision-making. The report concludes that the challenge of better pest and disease management in a rapidly changing world is potentially an invitation to devise more participatory, integrated frameworks for decision-making and priority-setting for agricultural research and development (Pretty & Bharucha, 2015). More broadly, it is an invitation for those engaged in challenging agriculture to turn their energies to the role of integration as the key to mutual learning and change.

Solving the growing distance in pest management strategic and tactical choices policy and research can no longer be the sole concern and responsibility of a single agency, organization or institution. Rather it must now become a common and collective concern and responsibility across national and regional stakeholders of agriculture who both define, share and ultimately negotiate an explicit framework for choice and action (White & Lazarus, 1986). Globalization, trade liberalization, technological developments, competitive behavior in commodity and input markets and growing consumer concerns on health and environmental safety are rapidly changing pest management concerns, threats and challenges for food and fiber producers and other stakeholders regardless of where they sit, be it in a developed or developing country, in the North and in the South. Unless agricultural stakeholders form and maintain inclusive alliances for collaboration it is unlikely they can coherently or effectively address actions and countermeasures. Coupled with the slide of agriculture in many countries and increased arm's length arrangements governing commodity chain operations the significance and effectiveness of pest management options unrelated to stakeholder interests and understandings is diminished. Ultimately it is agricultural stakeholders and communities who will determine the limits of policy and research tolerance, affordance and support for pesticide use, and the future of agriculture and food for the Asia-Pacific region.

10.1. Case Study 1: Region A

The focus of this case study is on two West African countries gaining new potato varieties. The case study results represent formative evaluations that prompted changes in future analytical work. Potato is gaining importance in West and Central Africa due to its advantages over other staple crops, like grain and root crops. Research and development partners created a long-term strategy for potato development in West and Central Africa. The results from the first phase of region seeds and partners started close collaboration to develop these technologies. Scientists planned a methodological research agenda emphasizing participatory monitoring and evaluation.

With regard to coverage (expected outputs), it became clear that surveys needed to be commissioned on the dissemination of the new varieties. Also, some expectation on methods (with respect to logistics) were not realistic. With regard to sampling frame and strategy, two separate sampling frames were needed (farmer experimenting with the new varieties needs lifting potato plots to be spatially stratified). Impacts were envisaged an a range of impacts, with baseline values for many of them provided.

The impact of adoption on production, input use, and livelihoods was to be disaggregated by gender. The coverage of field observations of new variety plots was discussed (it was suggested that this be reduced). Coverage with respect to control plots was also discussed (after long deliberation it was decided a judicial exploitation of the situation in control plotting was sufficient). Efforts were taken to ensure that few measures and observations were too general. A computing exercise was suggested to reduce the number of variables in a way that retained as much information as possible and proved to have a high statistical power. Stressing purpose-directed disaggregation and only collecting relevant information on the process was also considered. Sensitivity analysis had not, as yet, been explored in depth. It was agreed that a statistical power analysis would take place, but this was seen as secondary to ensuring full coverage and clear definitions of endogenous control, pre-tests coordination.

10.2. Case Study 2: Region B

Well-established systems to routinely monitor, evaluate, and report pesticide use in accordance with the integrated pest management strategies and tactics were developed and incorporated into the overall programs. These monitoring and reporting systems included the following components:

(1) data on the most important pests and parasites, their respective economic thresholds, estimated distribution, cost and efficacy of their control, and use, and availability and ability of the control agents;

(2) estimates of pesticide use on all control agents in juvenile and mature production;

(3) a weekly summary of sampled inspection data collected, and state-wide estimates of the performance of those developments;

(4) documentation of implementation and adoption of each action arrested, including initial pest observations, reports of control agents and individual parameters on their use; and

(5) improvements in pest control or losses from pest-related damage.

In cooperation with staff, the Cooperative Extension evaluated the overall effectiveness of programs via a statewide survey. Since then, performance measures have been weekly used to evaluate recent commodity-oriented programs. At the prompting of improvement in field scouting alone was sought in maturity indices, a sensitive PFR, and a description and evaluation of harvesting machinery. Laboratory methods for diagnosis of virus were validated and integrated into the successful monitoring of control measures. With the realization of invertebrate Pest Management applications there is an ever pressing need for the understanding of software programs, and on-farm loop monitoring, or the concurrent evaluation of systems. But these programs need to be set up within a multi-national framework of existing long-term records in a form that will provide critical comparisons between widely different climates and farming conditions.

10.3. Case Study 3: Region C

Region C is a tourist area in the southeast with 12% of the population self-identified as Black. Four case farms varied from 25 to 200 acres irrigation systems. Wives of farmers served as project leaders. Farm visits occurred in the planning stage and in April each year. Following the April farm visits, a list of pest problems and activity plans were prepared for the case farms. Each farmer received a forecast of pest problems drawn from insect calendar graphs. Follow-up telephone calls occurred to prepare for each project visit. Cumulative pest data and pest treatment decisions were recorded every week. The intention was to show trends in population development, demonstrate that a decision was rational, and provide a record to evaluate the response of the pest to treatments (White & Lazarus, 1986). Farmers decided on treatments and methods. Home-made traps were produced and supplied to case farms. The goal was mosquito surveys on honeydew production and aphid population research, which impacted treatment schedules. The producers increasingly solved pest problems independently as the project progressed. Full responsibility was taken for execution of treatments.

Records were mostly written on computer in the last year. Farmers initiated pre-pest activity plan requests from cooperating agents. A community meeting discussing the project occurred at the annual producers' meeting. Reports to growers were hand-written and photocopied in the first few years. The last report is usually in colored prints. Painting, checkout, and calendar data were disseminated through a little information pamphlet. Research reports were in journal articles, newsletters, question-answer sheets, and reports to the agency and funding organizations. An overall decrease in the frequency of pest loss with greater use of non-chemical control options coincided with the changes in farmers' treatment decisions. Accounting for variations in interpretations led to uncertainties in understanding the data. Inferences from monitoring data may be valid even if the monitored outcome is not measured.

11. Future Directions for MERL in IPM

Future directions for Monitoring, Evaluation, Research, and Learning (MERL) in Integrated Pest Management (IPM) can significantly enhance the effectiveness of pest management practices and promote sustainable agricultural methods. Some potential areas of focus include:

1. **Data-Driven Decision Making**: Utilizing big data analytics and satellite imagery for informed pest management strategies can optimize pest control measures.
2. **Integration of Technology**: The application of advanced technologies such as drones and IoT devices in monitoring pest populations can provide real-time data, improving the precision of interventions.
3. **Stakeholder Engagement**: Collaboration among farmers, researchers, and policymakers is crucial for developing inclusive IPM programs that cater to the needs of agricultural actors.

4. **Sustainability Assessment**: Creating frameworks to evaluate the environmental, economic, and social impacts of IPM practices will help in assessing their sustainability.
5. **Education and Training**: Continuous education on emerging IPM techniques for farmers can enhance the adoption of sustainable practices.
6. **Policy Development**: Advocating for supportive policies that facilitate research and funding for IPM initiatives can foster an environment conducive to innovation.
7. **Adoption of Biocontrol Agents**: Encouraging research into biological control agents as alternatives to chemical pesticides can lead to more sustainable pest management solutions.
8. **Climate Change Considerations**: Focusing on the impacts of climate change on pest dynamics can help adapt pest management strategies accordingly.
9. **Field Trials and Research**: Conducting field trials to evaluate the effectiveness of various IPM strategies will provide practical insights for refining pest management practices.
10. **Global Collaboration**: Promoting international collaboration in pest management research can lead to innovative solutions and a broader understanding of challenges.

Each of these directions emphasizes the need for a comprehensive approach to enhance the adoption and efficiency of IPM practices, aligning with both current agricultural needs and future challenges.

11.1. Emerging Technologies in Data Collection

The use of innovative technologies to garner and communicate data is rapidly expanding across multiple sectors and countries. In recent years, practitioners of the Monitoring, Evaluation, Research, and Learning (MERL) and research discipline have begun to explore how to harness these technologies to improve theory of change testing, research quality, and efficiency of information gathering and communication. Although the potential benefits of taking such an approach for MERL within the agriculture and food security sector have been recognized, there has been insufficient reflection to date on emerging technologies in MERL in agriculture and food security projects. This gap is problematic because it may result in missed opportunities for agriculture and food security projects to improve MERL via innovative technologies.

Discussions with leading experts in MERL in diverse agriculture and food security projects, including integrated pest management projects, have brought forth insights about emerging technologies with the highest relevance for effective data collection in agriculture and food security projects; barriers and facilitative factors to their adoption and continued use; and priorities for addressing capacity development needs to facilitate the responsible deployment of emerging technologies for data collection in agriculture and food security projects (Centre for Agricultural and Rural Cooperation, 2015). The identified technologies in the agriculture and food security sector included participatory citizen science and crowdsourcing; self-reporting apps; and low-cost easy-to-implement computer vision solutions. Existing barriers were a lack of awareness of new technologies; reluctance to change; insufficient integration into project procedures; a lack of human capacity and institutions; limited technological infrastructure; and poorly functional, low-cost technologies. By way of contrast, suggested facilitative factors to support their continued use included project and institutional stability; national integration; familiarity or experience with technology use; strong personal motivation and belief; and high youth population.

11.2. Innovative Approaches to Impact Assessment

In Indonesia, Participatory Learning and Action (PLA) methods are being used to assess the impact of the Improved Water and Soil Conservation technology packages developed by the Integrated Catchment Management for Upper Citarum Project. Lessons learned in using Learning History and Impact Pathway Development tools, developing and applying impact assessment questions, negotiating to balance interests of multiple stakeholders, and using the community's social fabric for impact assessment coverage are shared. Since its first stakeholder consultation in December 2009, impact assessments using PLA approaches have increasingly been accepted by governmental and non-governmental organizations working at upper Citarum for a clearer understanding of the impacts. PLA is considered a relevant method for impact assessment as it enables effective engagement particularly with farmers. In HABITAT, a new approach is taken to evaluating and reviewing efficacy and impact. The new approach builds on the successful elements of the past but does not case the internal and external review in terms of group workshops. Instead, pilot testing of new tools and their application on a remote or virtual basis is undertaken in evaluation walkthroughs, adjusting to the current restrictions on international travel.

The PLA approach draws on hindsight, foresight and insight-based methods in Learning History, Impact Pathway Development, and Scenario Development respectively, including participatory goal setting through agreed pathways, and prioritizing steps of action collaboratively. Through consideration of multiple, internal and external perspectives on impact assessment, a more comprehensive, action-oriented, socially sensitive and feasible methodology that includes systemic input is derived. The methodology's implementation has taken place in the UCI Project, using extracted PLAs to assess technical impacts, commitment-based participation, and impact paths. Aiming to validate the implementation locally and community-oriented in Wastareja Watershed, the methodology reproducibly with the farmers as actors through research capacity building has been applied in all Australian Federal Government-supported project sites in Indonesia. Findings reveal multi-dimensions of active engagement through residential working in the same social and geographical environment, milestone commitments, adaptation on multi-levels, detectable and undetectable impacts, as well as directional, temporal, and spatio-temporal variations on development paths (Wienarto et al., 2016).

12. Conclusion

This MERL strategy is intended to remain a living document. The goal in December 2020 at the completion of this strategy will be to lay out a clear guideline for what success looks like in sensing, understanding, ceasing or reversing loss of biodiversity and how those senses might be facilitated. While care has been used to identify and analyze what lies ahead in this document, it may have some gaps and blind spots. Project leaders and MERL team are encouraged to add new ideas that emerge as the project advances. Also, now is the least appropriate time to focus on getting the details perfectly defined. The Melissa team must collectively and iteratively build up a nuanced understanding about biodiversity loss, influencing factors and their interconnections. MERL team and project leaders are encouraged to treat this strategy as a foundation for ongoing collaborative efforts to ensure the project is directing efforts towards the best possible outcomes. Finally, it is important to keep updated with extension impacts of other MERL teams. They can be invaluable inspirations as this team works to shape your project MERL strategy. Six core guide questions for monitoring and evaluation need to be elaborated below for each indicator and how to know if it is being successful. Each indicator can be adapted to be applied across focal contexts. The contextual application research cluster could be further elaborated or replaced with hopes.

An emergent outcome of this project is to collectively develop a relevant and actionable MERL framework. While all projects must lay out a MERL strategy for annual outcomes reporting, the Melissa project must enable relevant and actionable MERL to achieve its goal. Those outcomes need to be a lucid guideline on monitoring, evaluating and getting feedback on research outcomes (what works, for whom, where and why?). This guideline must enable ongoing iterative improvement of the settings, instruments and implementation of research. Measuring and communicating about success is challenging. It must change behaviours or a quality that is important for human well-being. Measuring current wellbeing does not indicate that long-term trends are improving or deteriorating (Pretty & Pervez Bharucha, 2015). Impacts (outcomes and possible unintended consequences) are difficult to observe because they happen over years. There are many external factors that influence outcomes and how these factors interact is poorly understood (Parsa et al., 2014).

References:

Almeida, F., 2018. Strategies to Perform a Mixed Methods Study. [PDF]

Avila, E. M., L. Tolentino, L., B. Binondo, C., L. Perez, M., & Marina Apgar, J., 2016. Building Bridges to an Uncertain Future Lived Now: Lessons from the Use of Participatory Action Research and Theory of Change Towards A Realistic Community-Based Participatory Monitoring and Evaluation System. [PDF]

Barr, D. B., Landsittel, D., Nishioka, M., Thomas, K., Curwin, B., Raymer, J., C. Donnelly, K., McCauley, L., & Barry Ryan, P., 2006. A Survey of Laboratory and Statistical Issues Related to Farmworker Exposure Studies. ncbi.nlm.nih.gov

Boye, K., Lindström, B., Boström, G., & Kreuger, J., 2019. Long-term data from the Swedish national environmental monitoring program of pesticides in surface waters. [PDF]

Capegsan, C. B., 2019. Participatory monitoring and evaluation in rural development: Capitalizing on the CHARM2 experience. [PDF]

Centre for Agricultural and Rural Cooperation, 2015. MEAgriSys. [PDF]

Centre for Agricultural and Rural Cooperation, T., 2015. Q&A: The status of ICTs in integrated pest management. [PDF]

Clark, L. & Marina Apgar, J., 2019. Unpacking the Impact of International Development: Resource Guide 4. Developing a MEL Approach. [PDF]

Derr, T., Moote, M. A., Savage, M., Schumann, M., Abrams, J. B., McCarthy, L., & Lowe, K., 2005. Monitoring Handbook 5: Monitoring social and economic effects of forest restoration. [PDF]

de Oliveira Lima, D.C., Alves Ramos, M., Costa Hermenegildo da Silva, H., & Giuseppe Chaves Alves, A., 2016. Rapid assessment of insect fauna based on local knowledge: comparing ecological and ethnobiological methods. ncbi.nlm.nih.gov

Garley, A., Eckert, E., Sie, A., Ye, M., Malm, K., A. Afari, E., Sawadogo, M., Herrera, S., Ivanovich, E., & Ye, Y., 2016. Strengthening individual capacity in monitoring and evaluation of malaria control programmes to streamline M&E systems and enhance information use in malaria endemic countries. ncbi.nlm.nih.gov

Hochachka, G. & Thomson, S., 2009. Developing capacity and community well-being : action research on an integral capacity development approach in the Mapacho River Watershed, Peru. [PDF]

Kosmowski, F., Stevenson, J., Campbell, J., Ambel, A., & Haile Tsegay, A., 2017. On the Ground or in the Air? A Methodological Experiment on Crop Residue Cover Measurement in Ethiopia. ncbi.nlm.nih.gov

Kumar, R. & Martin, A., 2017. Conceputalising, achieving and measuring impact. [PDF]

Lyman, S. E., 2019. Uncharted Territory: UVM Extension Data Visualization Needs Assessment. [PDF]

Matten, S. M., L. Hellmich, R., & Reynolds, A., 2004. Current resistance management strategies for Bt corn in the United States. [PDF]

Muleme, J., Kankya, C., C. Ssempebwa, J., Mazeri, S., & Muwonge, A., 2017. A Framework for Integrating Qualitative and Quantitative Data in Knowledge, Attitude, and Practice Studies: A Case Study of Pesticide Usage in Eastern Uganda. ncbi.nlm.nih.gov

Muriithi, B. V., D Affognon, H., M Diiro, G., W Kingori, S., M Tanga, C., W Nderitu, P., A Mohamed, S., & Ekesi, S., 2016. Impact assessment of Integrated Pest Management (IPM) strategy for suppression of mango-infesting fruit flies in Kenya. [PDF]

Parsa, S., Morse, S., Bonifacio, A., C B Chancellor, T., Condori, B., Crespo-Pérez, V., L A Hobbs, S., Kroschel, J., N Ba, M., Rebaudo, F., G Sherwood, S., J Vanek, S., Faye, E., A Herrera, M., & Dangles, O., 2014. Obstacles to integrated pest management adoption in developing countries. [PDF]

Pretty, J. & Pervez Bharucha, Z., 2015. Integrated Pest Management for Sustainable Intensification of Agriculture in Asia and Africa. ncbi.nlm.nih.gov

Schuetz, T., Meadu, V., Atakos, V., M. White, J., Schubert, C., Luis Urrea Benitez, J., L. Ampaire, E., A.O. Radeny, M., Mungai, C., & Cramer, L., 2017. Measuring how communication and engagement efforts help deliver outcomes. [PDF]

Sokol-Oxman, L., 2015. Implementation of a Participatory Approach to Monitoring and Evaluation: Literature Review & Case Study Application. [PDF]

Star, P., 1999. Data Analysis. [PDF]

White, G. B. & S. Lazarus, S., 1986. TB116: Integrated Systems for Managing Potatoes in the Northeast. [PDF]

Wienarto, N., Kuswara, E., & Hakim, A., 2016. Picturing impact of the PEDIGREA program: a case study from Indramayu, Indonesia. [PDF]

Yang, X., D. Blower, J., Bastin, L., Lush, V., Zabala, A., Masó, J., Cornford, D., Díaz, P., & Lumsden, J., 2013. An integrated view of data quality in Earth observation. ncbi.nlm.nih.gov

Kolibri 360 GmbH
Ferdinand-Jühlke-Straße 7
99095 Erfurt, Deutschland
produktsicherheit@kolibri360.de

Druck:
CPI Druckdienstleistungen GmbH
im Auftrag der
Kolibri 360 GmbH
Ferdinand-Jühlke-Str. 7
99095 Erfurt